WHO PROVIDES INSPECTION

REPORT BY THE ASCE CONSTRUCTION DIVISION INSPECTION COMMITTEE

EDITED BY
Abdol R. Chini

COMMITTEE MEMBERS
I. Lloyd Guthrie, David M. Hummel, Jr.
Stuart H. Williams, Thomas S. Wolf

TECHNICAL REVIEWERS
Cliff Schexnayder, Jeffrey L. Groom, Raymond Hays
Roland S. Jary, Thomas R. Rath, James A. Tomasulo

ASCE American Society of Civil Engineers
1801 ALEXANDER BELL DRIVE
RESTON, VIRGINIA 20191–4400

Abstract: With the increasing interest in the design-build project delivery system, the decision of who should be responsible for quality assurance/quality control has become a complex one. To assist owners in this decision, ASCE's Construction Division Inspection Committee developed the monograph, *Who Provides Inspection.* The monograph is concerned with the advantages and disadvantages of methods used to perform and administer various quality assurance/quality control programs for different types of construction. Each of the papers deals with a different segment of the construction industry: heavy civil (dams and waterways), transportation, water and wastewater, commercial buildings, residential buildings, and waste management.

Library of Congress Cataloging-in-Publication Data

Who provides inspection: report / ASCE Inspection Committee; edited by Abdol R. Chini.

p. cm.

Includes bibliographical references and index.

ISBN 0-7844-0450-X

1. Engineering inspection. 2. Civil engineering--Quality control. 3. Quality assurance. I. Chini, Abdol R. II. American Society of Civil Engineers. Construction Division. Committee on Inspection.

TA191.W46 1999

624'.0068'5--dc21 99-33479

CIP

Library of Congress Catalog Card No: 99-33479 ISBN 0-7844-0450-X
Manufactured in the United States of America.

Foreword

On a given project, construction Quality Assurance/Quality Control (QA/QC) programs may be administered by the architect-engineer, contractor, special consultant or project owner. One of the most important questions owners must consider in the planning stage of any construction project is, "who should provide inspection services on the project?"

To clear up some of the confusion in the process as it exists today, ASCE's Inspection Committee has developed this monograph. It is concerned with the advantages and disadvantages of methods used to perform and administer various QA/QC programs for different types of construction. Each of the papers included in the monograph deals with a different segment of the construction industry: heavy civil (dams and waterways), transportation, water and wastewater, commercial buildings, residential building, and waste management.

The Committee wants to express its sincere appreciation to Construction Division EXCOM member Cliff Schexnayder who reviewed the original document and offered many helpful suggestions. Thanks also to the following reviewers for their suggestions: Jeffrey L. Groom, Commercial Testing Laboratories; Raymond Hays, Rust Engineering & Construction; Roland S. Jary of Fort Worth, TX; Thomas R. Rath, Silverado Constructors; and James A. Tomasulo, San Diego Metropolitan Wastewater Department. Finally, the following Inspection Committee members contributed extensively in production and publication of this monograph: I. Lloyd Guthrie , Stone Webster Engineering Corp.; David M. Hummel, Jr., Maxim Technologies, Inc.; Stuart H. Williams of Culver City, CA; and Thomas S. Wolf, CH2M Hill.

Abdol R. Chini
Editor

Contents

WHO PROVIDES INSPECTION

Construction Division's Committee on Inspection

Not that the question was ever an easy one to answer, but in this time of increasing interest in Design-Build project delivery system, the question of "Who provides inspection?" has become even more complex. The "inspector" alternatives have always been clear:

- Designer
- Owner
- Independent Agency
- Contractor

Each of these alternatives is viable, and, for any given set of project parameters, there may be a clear, rational argument for one being the best. Whatever the choice, it is important to know that there are choices, and that the best choice will most likely require deliberate and careful consideration.

This monograph presents and discusses the pros and cons of each choice within the context of the following fields of construction:

- Heavy Civil - Dams and Waterways
- Commercial Construction
- Water/Wastewater
- Residential
- Waste Management
- Transportation

In the committee's opinion, there is no single "best choice" inspection alternative for a given field of construction. The inspector decision will be based on a combination of both objective and subjective criteria, and it could be that there is no latitude for consideration of alternatives, as in the case of many US Government projects that require the Contractor to provide the inspection.

The committee also believes that there are a number of fundamental truths, or paradigms, that should be understood and considered in the "who provides inspection" decision. In its final committee meeting on this subject, the committee developed the following list for your consideration:

- Quality and inspection are directly related.

- Quality is a team activity.

- Quality must be in the specification. More cannot be expected without associated additional costs.

- The designer of record, if involved in inspection, should be informed of this in the original design scope of service.

- Inspections should be discussed and planned during design development and specification drafting. Engineers Joint Contract Documents Committee (EJCDC) provides helpful guidance.

- Partnering can improve inspection communications and reduce the potential for confrontation.

- Qualifications (accreditation for laboratories and certifications for inspectors) must be specified and appropriate for the type of construction.

- Laboratories and Inspectors must be independent.

- A Professional Engineer must have responsibility for verification of test results and certification of completed project.

- The Owner's role in the QA/QC process is to set costs and define expectations.

- When properly specified and performed, inspection verifies construction quality and satisfaction of design intent.

The following papers present the thoughts of committee members on the subject of "Who Provides Inspection". Consideration of this information will hopefully make the next decision on who provides inspection better informed and the next project more successful.

QUALITY ASSURANCE/QUALITY CONTROL IN THE CONSTRUCTED PROJECT

Abdol R. Chini, Ph.D., P.E.,[1] Member, ASCE

INTRODUCTION

> "Quality in the constructed project is much more than merely quality assurance and quality control (QA/QC) programs carried out by the project participants. It is a state of mind of all involved in the project that places quality foremost" (ASCE, 90).

Competent and objective inspection is mandatory if owners are to be assured that their projects will conform with the contract requirements and ultimately become quality products. The main reason for the need of QA/QC programs is to be able to locate or trace the cause of construction faults. Another reason is that although there has been improved developments in management and technology in the construction industry, the quality standards have not been improved proportionally. As a result, faulty designs and construction have caused numerous damages and loss of life and property.

QA/QC should be the responsibility of the entire project team, including the owner. The owner and designer must fully understand the issues of costs and tolerances inherent to QA/QC and recognize that absolute perfection is beyond the owner's budget. This is where successful partnering can level the understanding of QA/QC among all participants. Quality control inspections should be done with the motive of encouraging and ensuring good workmanship rather than catching the culprits. The major cause of conflicts pertaining to QA/QC programs is the delegation of responsibility and authority for providing inspection, determining acceptance and enforcing compliance.

During the design phase, architect-engineers are solely responsible for quality of the design. But, during construction, it is the responsibility of the contractors that construction conform to the quality standards in the contract documents. On a given project, the QA/QC programs may be administered by the architect-engineer, contractor, special consultants or the project owner. Every project is uniquely different from the next, so there is no set standard in selecting which approach is more effective. This paper is concerned with the advantages and disadvantages of each method concerning responsibility and authority in performing and administering QA/QC programs in different types of construction.

---

[1] Associate Professor, Rinker School of Building Construction, University of Florida, Gainesville, FL 32611.

It should be noted that while inspecting, testing, and monitoring to evaluate materials used in construction may be performed by qualified technicians, they should be performed under the technical supervision of a licensed engineer, insofar as it involves safeguarding life, health, or property. Not that license makes engineers smarter, but because they are accountable to the laws of the state where they practice.

Not discussed in this paper is the role of the independent materials inspection and testing firms and who should employ them. These firms provide valuable services to the construction industry and owners by determining engineering properties, interpreting data, evaluating and monitoring the materials, products and processes used in construction.

In recent years the construction industry has turned its attention to the implementation of Total Quality Management (TQM). TQM is a management approach that attempts to focus more on problem solving as opposed to traditional approach of monitoring and inspection. In addition, ISO 9000 is lately receiving much attention in construction industry. Certification to ISO 9000 confirms that the company has been audited by a certifying organization and its practices and procedures are consistent with ISO 9000 Standards. These two new approaches in quality assurance will be reviewed later in this paper.

DEFINITIONS

"Quality assurance is defined as the plan to effect the performance of design and construction" (Goldberg, 91). Quality assurance programs should encompass the establishment of the procedure for defining, developing and determining quality standards in design and construction. Another requirement is developing the procedures to monitor, test, inspect, measure, document and review activities to assure compliance with established quality standards in regard to construction materials, methods and personnel. Defining the administrative procedures and requirements, organizational relationship and responsibilities, to assure attainment of the established quality standards is also an important aspect of this definition.

Quality control is that element of the quality assurance program which entails performing inspections, testing and measurement, and documentation necessary to ensure conformance with the contract specifications on the quality of construction materials and methods. "Quality control is defined as the method to monitor the results of the quality assurance plan" (Goldberg, 91). A quality control program should include a definition for the standard of quality for a particular project and a defined procedure to measure achieving such standard. Also such program needs to define how to execute the procedure of measurement to determine whether the standard of quality was attained. Ultimately, it should define the power to enforce and maintain the defined standard of quality as measured by the specified procedure.

RESPONSIBILITY AND AUTHORITY IN QA/QC

Four common approaches can be identified regarding the responsibility and authority to perform and administer a QA/QC program in a construction project. The responsible party for the QA/QC program can be the architect-engineer, contractor, independent special consultant, or the project owner with in-house QA/QC personnel.

Architect-Engineer

Due to the increase in liability claims against design professionals, particularly architects, the American Institute of Architects standard form of agreement between the architect and the owner (AIA, Document A 201) has been modified. It indicates that architects are expected to make only periodic on-site observations, and they are not expected to make exhaustive on-site inspection to check the quantity or quality of the work unless otherwise agreed upon (Fisk, 92). If designs are technically complex or require frequent interpretation during construction, it would be beneficial for architects to develop and administer their own QA/QC program to assist the construction. The designers field staff would perform necessary testing and inspections, administrative and surveillance services, and coordinate specialized consultants to perform field and laboratory testing and inspections.

As a result this will provide prompt resolution of design disputes and implement efficient project communication, as well as, eliminate the problems of understanding the design standards and procedures. The owner and designer will also economically benefit on the management and liability costs associated with architect-engineer firms providing their own QA/QC program.

One of the two major drawbacks to this approach is that many architect-engineer firms are not adequately staffed to implement QA/QC services efficiently. The other concern is that if plan errors occur, the architect-engineer firms may be inclined to ignore proper procedure to cover up their own mistakes.

Contractor

Some public and private organizations have required prime contractors to take a more active role in the control of project quality by making them develop and manage their own QA/QC programs (Iffland, 88). The reason for requesting contractors to be responsible for QA/QC programs is that contractors have contractual obligations to produce the quality specified in the contract documents and should therefore, be responsible for the quality of their work. Contractors would be responsible for performing testing and inspection to control the quality of the work. They could either use their own manpower or subcontract all or part of the testing, inspection and engineering functions to testing laboratories and consulting engineers. They also would perform acceptance inspection and testing to verify conformance to quality standards and prepare testing and inspection documents.

In the case of contractor provided QA/QC programs, the cost of the program implementation to the owner will be reduced. Another benefit is the elimination of delays, because the contractor will not have to wait for inspections and approvals of the work by the owner or his/her agent (ASCE, 90). The most important advantage is that the contractor will be more responsive to prevent rework and replacement costs.

A few disadvantages to this approach must be noted, such as, the increased cost to the contractor to operate a QA/QC program and to handle the elaborate documentation and inspection requirements. Also the QA/QC program's effectiveness depends on the contractor's capability to organize and manage his/her staff to meet QA/QC objectives. Conflicts between the production team and the QA/QC personnel in execution of the QA/QC program can surface as a disadvantage to this approach.

Special Consultant

Owners sometimes hire special consultants that are not related to the project designer, to provide all or part of the QA/QC services. Such consultants include construction management firms, special testing laboratories, and specialized consulting engineers (Clough, 86). They can be involved in developing contract requirements and documents for QA/QC programs of preconstruction stages and perform or supervise QA/QC functions during construction.

The overriding advantage in the use of independent special consultants is their loyalty to the owner, since the consultant only has a contractual relationship with the owner, not the designer or contractor. This in turn would reduce the potential conflict of interest and collusion between the designer and contractor. Another benefit to this arrangement is that the independent special consultant firms are competently better equipped and staffed to handle QA/QC programs than the contractor or designer (O'Brien, 89). Finally, the design fees accrued by the architect-engineer would be reduced due to the fact that no full-time supervision is required of the project designer.

Unfavorable situations may occur with this setup, for instance, the independent special consultant's late involvement in the design process reduces his/her early familiarity with design criteria and quality requirements. This will restrain their ability to develop and implement efficient and appropriate QA/QC programs. This problem can be minimized if consultants are involved in establishing project specifications early in the design phase. Problems often arise concerning responsibility and authority regarding some contract administrative functions. This could create adversary relationships between the designer and the special consultants which would put a restraint on project communication. This situation could be avoided if the responsibilities and functions of the various participants in the construction process are explicitly defined in the contract documents.

Owner

Finally, government agencies and corporate and institutional owners who are continually in the construction market may find it necessary to have an in-house project management staff whose responsibilities will include administering QA/QC programs. The effectiveness of such force account programs depends on the availability of well-qualified field personnel to monitor and administer QA/QC programs efficiently.

An advantage of the project owner using his own force account is having adequately trained personnel within the organization who can reduce the time and cost due to familiarity with individual project QA/QC programs. Another aid to this approach is the better control of field supervision and direct control of quality.

To offset these advantages there are some disadvantages to this approach. For instance, to engage in a full-time staff, unless there is always a project going on, would be uneconomical to the owner. The other disadvantage is that organizations with force account generally have established standards and quality assurance procedures in which their staffs are trained. This greatly limits flexibility and suggestions to change in construction methods, materials, quality requirements and test procedures. However, this can be minimized by organizing continuing education programs for the staff and updating QA/QC procedures to meet current developments in the industry. Adequate supervision of field staff is necessary to ensure compliance to new techniques.

TOTAL QUALITY MANAGEMENT

In recent years the construction industry has turned its attention to the implementation of Total Quality Management (TQM). TQM is a management approach that attempts to focus more on problem solving as opposed to the traditional approach of monitoring and inspecting. Organizations embracing TQM are adopting a management philosophy that makes quality a strategic objective for the organization (Total, 1990). The focus is to improve a process through elimination or reduction of defects and low value tasks. The foundations of TQM rest on the individual workers and their participation. Managers must create the proper environment for the workers to contribute.

TQM emphasizes customer satisfaction and continuous improvement (Chase, 1993). Within the construction industry, each party involved on a project, including the owner, contractor, and designer, plays the role of customer and supplier of services. The owner supplies the requirements to the designer, the designer supplies the plans and specifications to the contractor, and the contractor supplies the built facility to the owner (Juran, 1988). A principle focus of TQM is for each supplier of services to identify and satisfy or exceed their customer's needs in terms of cost, quality, and time.

Construction has lagged behind other industries in implementing TQM. Part of the reason is the unique characteristics of construction projects , such as short duration, transient labor force, and the fact that each project is somewhat different. Another part is that instead of focusing on processes and causes of problems, the industry has focused on monitoring and inspection of results. For a successful application of TQM in construction, three major focuses need to be addressed. These include focus on detailed needs of customer, integration of the various technical skills and management know-how of the diverse parties involved in the construction process into a common approach to problem solving, and focus on investigation of causes of problems.

ISO 9000

One quality management system receiving much attention in the construction industry lately is ISO 9000. ISO 9000 refer to a series of international quality management and assurance standards that provide a benchmark for compliance. These standards apply to all services as well as the organizational procedures which contribute to the quality of the end product or service. ISO 9000 standards were issued and approved in 1987 by 35 countries (Marquardt, 1994). Today, more than 100 countries, including the U.S., Japan, and the European Union have adopted these standards as a national standard. The U.S. Department of Defense has approved use of ISO 9000 standards in lieu of MIL-Q-9858A whenever practical and the ASCE has adopted the policy statement #431 which explicitly supports a worldwide registration and recognition under the ISO 9000.

Certification to ISO 9000 confirms that the company has been audited by a certifying organization and its practices and procedures are consistent with ISO 9000 standards. The auditors compare the company's quality manual with the ISO 9000 standards and verify that procedures and systems are consistent with the manual. In other words, a firm is certified that they actually do what they say they do. The principal requirement for certification is documentation. A company has to document its business processes and show how these processes are followed and monitored for compliance with the firm's quality manual.

Since the European Community has adopted the ISO 9000 series as a model for quality management system, construction contractors in Europe are seeking and obtaining ISO certification. Important reasons they seek certification are customer demands, marketing advantages and desire to establish a functional quality management system. An added push for certification comes from those ISO certified firms that may require any organization doing business with them to be ISO certified. A 1994 listing of registered companies in the United Kingdom (DTI QA Register, 1994) included more than 500 listings for construction companies. These companies are nearly in all sectors of the commercial and industrial construction industry as well as engineering.

Trends toward certification in other industries could soon compel some US construction companies to seek certification. Many large domestic industrial manufacturers have pursued certification for access to European markets, and may begin requiring certification for contractors bidding on private construction projects (Fairweather, 1994). Currently, less than twenty firms with a Standard Industry Code in a construction division have US site registered to a specific ISO standard.

CONCLUSIONS AND RECOMMENDATIONS

Quality control should be the responsibility of the entire project team, including the owner. However, there is a need for a systematic control of quality in the construction process. All four approaches regarding the responsibility and authority to perform and administer QA/QC programs as outlined here offer advantages and disadvantages to both the provider and the recipient of the program. The most effective approach however, would have to be determined by the conditions of the proposed project.

REFERENCES

American Society of Civil Engineers (ASCE). (1990). *Quality in the Construction Project: A Guideline for Owners, Designers and Constructors* (Vol 1). New York: ASCE.

Chase, G. W. (1993). *Implementing TQM in a Construction Company,* The Associated General Contractors of America, Washington, DC.

Clough, Richard H. (1986). *Construction Contracting* (5th edition). New York: John Wiley & Sons Inc.

DTI QA Register. (1994). *The United Kingdom Register of Quality Assessed Companies,* supplement 2, HMSO, London, (pp 94-98).

Fairweather, V. (1994). "Understanding ISO 9000", Civil Engineering, Feb. 1994, (pp 59-61).

Fisk, Edward R. (1992). *Construction Project Administration* (4th Edition), New Jersey: Prentice Hall Inc.

Goldberg, S. W. (1991). Construction Inspection Responsibility. In Luh-Maan Chang (Ed.), *Preparing for Construction in the 21st Century* (pp 524-562). New York: ASCE.

Iffland, Jerome S. B. (1988, May). "Quality out of the Manual and into the Field", Civil Engineering (pp 54-56).

Juran, J. M. (1988). *Juran of Planning for Quality,* the Free Press, New York, NY.

Marquardt, D. W. (1994). "Background and Development of ISO 9000", *The ISO 9000 Handbook,* 2nd Ed., R. Peach Editor, CEEM Information Services, Fairfax, VA (pp 2-14).

O'Brien, James J. (1989). *Construction Inspection Handbook: Quality Assurance/Quality Control* (3rd Edition). New York: Van Nostrand Reinhold Publishing Co.

"Total Quality Management: The Competitive Edge", (1990). Publication 10-4, Const. Industry Inst., University of Texas at Austin, Austin, TX.

QUALITY AND CONSTRUCTION INSPECTION HEAVY CIVIL INDUSTRY: DAMS AND WATERWAYS

Robert H. Welsh, P.E.,[1] Member, ASCE
Richard A. Welsh, P.E.,[2] Member, ASCE

INTRODUCTION

The importance of construction inspection to the successful completion and long life of a project is well recognized in the construction industry. The importance of inspection becomes even more paramount to the successful completion of dams and water ways in the heavy civil industry due to the catastrophic consequences to life and property should a failure occur. Accordingly, both the level and degree of inspection in this segment of the heavy civil industry is usually higher than in some of the other segments of the industry. The options in use by the various owners, along with advantages and disadvantages will be discussed in this paper.

DEFINITIONS

For the purposes of this paper, dams and waterways are defined as large or major structures, i.e. dams of greater than 25 feet in hydraulic height with reservoirs of several hundred acre feet storage capacity; and canals or pipelines with capacities in excess of 100 cubic feet per second. Generally, structures of this magnitude are not governed by local building codes, nor can the owners rely on local authorities to provide the necessary level of inspection. (Note, these criteria were arbitrarily set by the authors and will not necessarily agree with criteria established by various owners in their definitions of major structures). Also, it should be noted that the authors have premised this paper on inspection being a part of the Quality Assurance Program as opposed to Quality Control Program.

The various owners of dams and waterways are generally in one of the following groups:

1. Governmental Agency

    Federal, such as Corps of Engineers, Bureau of Reclamation or Soil Conservation Service

---

[1]Construction Engineer, Bureau of Reclamation, Bend Construction Office, Bend, OR 97701.

[2]Field Engineer, Bureau of Reclamation, Willows Construction Office, Willows, CA 95988.

State, such as California Department of Water Resources

County or City, although not as common, there are examples such as Douglas County Oregon and the City of Seattle or the City of San Francisco.

2. Quasi Governmental Agency

Water Districts, such as Metropolitan Water District of Southern California or the Las Vegas Valley Water District in southern Nevada

Regional Water Agency/Authority, such as Placer County Water Agency in northern California or Tennessee Valley Authority.

Utility Districts, such as East Bay Municipal Utility District near San Francisco or Sacramento Municipal Utility District in northern California.

3. Private owners of dams and waterways such as privately owned power, water or mining companies.

OPTIONS

There are many options available to owners for providing inspection as part of quality assurance programs. A list of options available would include, but is not be limited to, the following:

1. Owner inspection using "in-house" inspection capability.

2. Inspection by a design professional (A&E firm). This could take two forms, either the design professional (A&E) who designed the facility or an A&E that is independent of the actual design A&E.

3. Inspection by a construction manager (CM), retained by the owner and independent from the design professional (A&E).

4. Inspection by the construction contractor, which would be a quality assurance function in addition to quality control responsibilities.

5. Inspection by independent testing laboratories, which could be separate or in addition to any quality control testing responsibilities the testing laboratories may have under the contract.

DISCUSSION

Of course, there are many combinations of the above that are or could be used by the various types of owners, depending on what best suits their needs and budget. With this in mind, a discussion of the various options, with potential advantages and disadvantages follows. This discussion is traditional in nature; i.e., it is recognized that there are shifts in the way of doing business currently taking place due to downsizing and shifting of priorities and changes in mission, especially within the larger Federal agencies, that may change how and by whom inspection is done.

Owner's "In House" Forces

Inspection of the construction of major dams and waterways by the owner's in-house staff is primarily found in the larger governmental agencies at the Federal and state level. Additionally, larger quasi governmental agencies, such as the Tennessee Valley Authority, make use of in-house inspection capability. Typically, owners who have the volume of work that supports in-house inspection capability, also have in-house design staffs, and could be characterized as design/construct agencies. Examples would include the Corps of Engineers, Bureau of Reclamation, Soil Conservation Service and Tennessee Valley Authority.

The degree of in-house capability will vary from owner to owner. Some owners staff the full range of quality assurance functions including survey, laboratory testing (materials control), and inspection. Other owners may contract separately for some functions or make some functions the responsibility of the construction contractor while maintaining responsibility for the inspection function.

Some of the advantages to this approach are:

a. You have a cadre of personnel who deal with similar plans, specifications, and designers if design is done in-house, so there is a high degree of consistency from project to project.

b. In general, and in theory, for the same level and degree of inspection, the cost should be less since the owner is not paying profit to a second or third party.

c. The owner has direct access and control over the inspection staff and can balance the inspection program to meet the agency's needs and budget allocated to the inspection program.

d. For those agencies who also have in-house design staffs, the design and inspection staff work closely together to insure the design intent is met.

Disadvantages would include:

a. The costs could very well be higher than other options unless there is a reasonably constant work load. This is one reason this option is most commonly used only by large owners who have relatively constant work loads. If the work load is not relatively constant, it will become uneconomical to retain the necessary staff.

b. Some will argue there is a potential conflict of interest when in-house personnel perform the inspection function.

Many owners do not have the workload volume that is necessary to maintain and justify an in-house design and/or quality assurance staff. For these owners, one of the other four options or combinations is the method of choice.

Independent Consultants

Probably, the most common would be inspection by consultant or construction manager (option nos. 2 or 3) or some combination, wherein independent consultants (A&E's primarily) are retained to provide design, quality assurance, or construction management services. There are many variables, from the scope and complexity of the project to the needs and budget of the owner which can influence the choice of methods to provide the inspection function.

An owner may choose to include the inspection function in the scope of services of the design A&E. The advantage of this is having a close working relationship between the designer and the inspection staff to insure design intent is met and the consistency derived by the inspectors working with familiar plans and specifications. The primary disadvantage is the potential conflict of interest issue (i.e., a tendency to cover up design errors or omissions) that is raised when the same firm designs and inspects.

A variation of the option of inspection by the design A&E is to have the inspection performed by an A&E that is independent from the design A&E. The advantages in this variation are that it can avoid the potential for a conflict of interest and it provides flexibility if the designer of choice does not provide inspection services. A potential disadvantage to this option, is that a more deliberate effort may have to be made to insure that a close working relationship is maintained between the designer and the inspection staff. In some cases this will require the scope of work in the design contract be expanded to include regular site visits and interfaces with the inspection team during the course of construction. Another potential disadvantage to this option is that an increase in the overall cost of the project is possible due to the multi-layering, overlapping of functions and increased coordination that is required. Typically, only the larger A&E firms will have personnel experienced in construction of dams and waterways.

Under the options just discussed it is assumed that the owner is performing the contract administration function. If the owner does not have this capability in-house then contract administration must be included in the scope of services of one of the A&E firms or the construction manager as discussed below.

Construction Manager

The next option, which is growing in use, is retaining the services of a construction manager to provide a full range of services including inspection. Typically, the CM is independent of the design professional (A&E) and the owner. Often the CM interfaces with the designer, the owner and the construction contractor. A full scope of services will include design and constructability reviews, preparation and issuance of the bid documents, conducting bid openings, awarding the contract, performing contract administration and performing the quality assurance functions including inspection. The scope of work however, can be limited to include only selected functions.

Advantages to this option are similar to those of using in-house capabilities except the CM employees do not work directly for the owner. There would be less layering and overlapping of functions under this option and it is suited to those owners who do not have contract administration capabilities in-house.

Disadvantages would include the fact that the owner must include a separate contract for the design services. (Note: it is recognized by the authors that a variation of the second option would be to retain an A&E that can provide the design services and a full range of CM services). In the Federal sector, it is not allowed to contract for the services of a Contracting Officer. Therefore, even if a Federal agency went with a CM, it would still have to perform certain functions in-house.

Contractor Forces

The fourth option, contractor inspection, although used, is not recognized as a good option within the heavy civil industry. Most reputable contractors agree it is not a desirable option because, while quality control is and should be the responsibility of the contractor, the quality assurance function should be the responsibility of the owner. Therefore, this option will not be discussed in any greater detail in this paper.

Independent Testing Firms or Laboratories

The final option, the use of an independent testing laboratory for the inspection, is not widely used in the heavy civil industry by and of itself. Typically independent testing laboratories are used in the testing and control of materials as part of the overall quality assurance program, but not for the inspection function in the construction of dams and waterways. The exceptions would be those larger firms that can afford to staff a cadre of personnel experienced in the construction and inspection of dams and water ways. Independent laboratories are often retained by A&E's or CM's as part of the quality assurance team on heavy civil projects.

CONCLUSIONS AND RECOMMENDATIONS

In conclusion, there are many ways to provide the inspection function in the construction of dams and waterways or the construction of other facilities found in the heavy civil industry. The method chosen will depend on the owner, the owner's needs and budget, and personal preferences. There is no one option that will be best for all owners and all situations.

Accordingly, the authors would like to recommend that owners weigh all of the variables before making a selection. However, no matter which option is chosen, we recommend the following:

1. That an adequate budget is provided to pay for the level of inspection commensurate with the structures being constructed and the risks associated with a failure, reduced life or increased maintenance of structures due to problems in construction.

2. That any option selected provide for a team that includes the owner, designer, inspectors and construction contractor working together to insure design intent is met and the quality of construction meets the owner and life cycle needs at a reasonable cost.

CONSTRUCTION INSPECTION IN THE TRANSPORTATION INDUSTRY

Stephen B. Quinn, P.E.,[1] Fellow, ASCE

INTRODUCTION

Historically, construction inspection in the transportation field has been more particular than in any other industry. Standard staffing involved in the transportation area generally is accomplished by a Resident Engineer, either licensed or with long-term experience in transportation construction, supplemented by a staff adequate to provide full-time construction inspection. This inspection is supplemented by having a testing laboratory on site under the direction of the Resident Engineer or on call at the discretion of the Resident Engineer to perform the required testing of materials utilized in the transportation project. Most, if not all, State Highway Departments have developed standardized frequency of tests as well as standardized tests. These tests closely parallel ASTM and/or AASHTO for the final approval of materials incorporated in the project. These tests range from soils (gradation, liquid and plastic limits, compaction factors, etc.), concrete (cylinders and beams for compression and flexure), cores, reinforcing steel, post-tensioning strands, pipe tests, and other materials incorporated in the finished product. This is defined as full service inspection. The municipalities have, to some degree, borrowed the standard of excellence established by the various Departments of Transportation around the country.

Full-service inspection can be accomplished in many different ways: owner inspection, design engineer inspection, independent design engineer inspection and contractor inspection. A relatively new method of constructing transportation facilities (to this country) is design-build. The country is now experimenting with some private toll roads. As design-build is still in the experimental stage, the comments of this paper will be confined to the standardized highway and bridge construction being utilized for the vast majority of U.S. transportation projects. Preliminary indications in the design-build industry is that the Contractor will perform the construction inspection as well as full quality control with the owner providing independent oversight.

Contractor quality control on any project is a must. The contractor is the only one physically as well as legally responsible for producing the completed project. In most contracts, the contractor is responsible for employing means and methods to accomplish the required construction. Contractors have long expressed the opinion that it is their intention to produce the best quality project they can within the contract proposal and according to the specifications which they have bid upon. They have also long felt that it is the owner's responsibility to provide the verification

[1] A Fellow and Vice President of HNTB Corporation.

testing related to the construction. The author strongly endorses the position of the contractor in this matter.

The relationship of the testing laboratory to the inspection process in the transportation industry has been established. The owner, either through its own laboratory or by contracting for laboratory services, performs the full requirements of testing as specified in the owner's frequency of tests and/or spelled out in the requirements of ASTM or AASHTO. Either method adequately serves the industry. Testing, whether it is performed by professional independent laboratories or the professional owner's laboratory, must conform to the ASTM/AASHTO's or Owner's Testing Methods. These methods are spelled out fully in the above noted standards. As a general rule, these requirements are handled by technicians.

The purpose of Construction Observation Services is to assure the owner that the work produced conforms to the intent of the plans and specifications. This is usually defined as generally accepted practices comparable to the normal standards of the industry. Perfection is strived for and defined as 100 percent conformance to the plans and specifications. The industry, however, is dependent on people who are not perfect. The plans are not perfect and contain some minimal errors where knowledge of the intent of the design is a requirement. The principal means of this quality assurance is by construction observation.

Following is a description of each method noted above for the inspection of construction and the advantages and disadvantages of each of the three methods evaluated. The three methods are Owner Inspection, Design Professional Inspection, and Independent Design professional Inspection. The author recommends that the Designer of Record inspect his own work. An acceptable alternate to this method is that the owner through its own professional staff (if the owner is not the Designer of Record) inspect the work. The author feels that the problems associated with an independent design professional can only serve to delay the construction process at the expense of the contractor and does not recommend this method.

Designer of Record Inspection

In the transportation field, the design of highway and bridge projects is traditionally done either one of two ways: the design is done in-house by the owner's design professional staff or by an independent consulting engineer. Most rail systems use some form of Construction Management System. Inspection under the Construction Management System is discussed elsewhere. Most transportation agencies have their own in-house professional design team. This applies, at the least, to most major municipalities (city and county governments) and to the State Departments of Transportation throughout the country. The agencies mentioned above will either perform the design services themselves or hire an outside consultant. As a general rule, an outside consultant is hired either because the complexity of the project is beyond the ability of the in-house design staff, or simply because the in-house design staff has sufficient work available and cannot handle the additional projects.

One of the primary benefits of construction inspection is to provide oversight to ensure the design requirements are met. Although major tragedies, such as the Hyatt Regency Hotel in Kansas City, MO and the 1987 L'Ambiance Plaza, are rare in the transportation industry, failures do occur. From an engineering perspective, most failures in construction resulting in death or serious injury involve underground construction such as trenching. There have, however, been a number of deaths associated with tunnel and bridge construction.

The principal benefit of Designer of Record inspection, whether the inspection be through the owner's staff or through contract being an agent of the owner, is the ability to achieve the intent of the design at the best cost to the owner. The best cost to the owner invariably involves the best cost to the contractor. When problems develop on a project, be it risk of the unexpected in construction or plan error or non-complying construction, the Designer of Record is responsible for recommending acceptance of the corrective action to the owner. The corrective action or changed conditions may or may not involve additional costs to the owner. The contractor on the project bids the project with the hope that the plans and specifications are perfect and there are no unexpected occurrences. When any deviations arise, decisions as to how to proceed have to be made in an expeditious fashion. These decisions, if they change design in any fashion, must involve the Designer of Record. If changes to the intent of the design take place without concurrence of the Designer of Record, the designer cannot assume the responsibility and the liability of the design. The contractor is interested in a quick and efficient resolution of any problems that may arise. When the designer inspects his own design, excellent lines of communication are available to provide quick and efficient solutions to construction problems. Construction and design personnel are on the same team as the owner's representative to obtain the best solution for all parties to the contract. If the Designer of Record is the owner, and like most municipal and state agencies has a standard inspection procedure, excellent consistency of results is achieved in the finished products.

There are some disadvantages to an independent Designer of Record's inspecting its own work. The independent design personnel may not be familiar with the owner's standards and basic flow of information. In addition, an independent design consultant may not be familiar with contracting procedures and may not have the personnel available on staff to inspect the work. These disadvantages generally are not problems when the owner is the Designer. Obviously, these problems can be overcome if it becomes the standard of the industry to have the Designer of Record inspect the work. If this becomes a requirement, then the independent Design Consultant would be in a position to effectively and efficiently retain an experienced staff to perform the work.

A number of owners strongly prefer not to have the Designer of Record inspect its own work. The principal concern expressed is that if plan errors occur, the Designer of Record is inclined to ignore proper procedure to cover up its own mistakes. It is the opinion of the author that this is not a valid complaint. The code of ethics of Professional Engineers would not permit the Designer to conceal its mistakes without notifying the owner. If, in fact, the owner feels that this is a possibility with the designer the owner has employed, the owner has probably employed the wrong designer.

Owner Inspection

Most owners in the transportation industry have their own inspection forces. The benefits derived from the agencies maintaining these forces are numerous. Primarily, they do not have to contract work out. They have personnel who are thoroughly familiar with their own standards, having worked on projects of the same nature. They are thoroughly trained in inspection and testing procedures and techniques and in the coordination of that inspection and testing. They are thoroughly familiar with the required documentation obtained and have developed standardized methods of producing a contract. There are disadvantages to the owner's utilizing his own forces for an independent designer's work. Agency personnel through their lengthy experience know the specifications and the desired results. Generally, through their vast experience, they also know how to handle problems. There is a strong tendency to handle problems as they arise without consulting the Designer of Record. This loses the owner some protection against faulty construction if the Construction Administration personnel make plan changes, or rule on solutions to the unexpected during construction. In addition, constantly being engaged in the same area with the same contractors leads to complacency and does not expose the owner's personnel to new techniques and/or new methods of construction in the industry.

Independent Design Consultant

Some agencies have employed an independent design consultant who has had no involvement in the design process. The owners who utilize this method state that the principal advantage is that they have independent review on site during the construction process. There is no tendency to cover up designer errors. In addition, an independent design consultant may bring a fresh approach to the project and, therefore, is in a position to recommend substantial changes in the design for the owner's benefit on overall project cost. The author does not recommend that this course of action be followed. The disadvantages in this type of system far outweigh any potential advantages that may accrue to the owner.

The independent design consultant generally brings the same disadvantages to a project that the Designer of Record may bring: unfamiliarity with the contract documents, unfamiliarity with the contract procedures, and unfamiliarity with the standards developed by the owner. In addition, the owner potentially could lose the liability protection he has from the Designer of Record if the independent design consultant permits deviations from the designer's plans.

The more important drawback of this system is obvious. As stated above, there is no such thing as a perfect set of contract documents. The Designer of Record is responsible for those contract documents which includes the plans and specifications and must sign off on any changes to those contract documents. The contractor, on the other hand, has bid a project with the intent of building the project as quickly and efficiently as possible. When problems arise, the contractor is entitled to quick resolution especially if those problems are caused by conflicts, conflicting plans and/or specifications. The independent design consultant, when discovering or being involved in these conflicts, must have them resolved by the Designer of Record. The process involved in proper communication and the documentation required for this process make any procedure developed a tedious one. The independent design consultant has a contract with the owner as does the Designer of Record. There is no contract between the two parties. The independent design consultant must, therefore, go through the owner for any clarification of the intent of the plans, contract documents or resolutions of problems with the contract documents. The owner must contact the Designer of Record for the resolution. The contractor, meanwhile is required to wait for direction from the inspecting agent. The independent design consultant, having no participation in the plan preparation process, does not know the intent of the design and is not familiar with the design. The consultant is placed in a position where he is required to interpret someone else's intent. In the opinion of the Society, this is a poor solution and should not be implemented.

CONCLUSIONS

Construction inspection in the Transportation Industry has been developed to ensure the public that highway and bridge facilities are safe. Methods of conducting the detailed inspection include Designer of Record Inspection, Owner Inspection, and Independent Design Consultant Inspection. The conclusion reached is the project can best be served by inspection being performed by the Designer of Record provided the Designer of Record has qualified personnel available. Owner Inspection where the owner employs qualified personnel is an acceptable alternative.

INSPECTION IN HIGHWAY/TRANSPORTATION CONSTRUCTION

R. Edward Minchin, Jr., P.E.,[1] Member, ASCE

INTRODUCTION

The construction of highways and other transportation facilities presents an array of challenges unique to that type work. The horizontal construction aspect, maintenance of traffic, the size and variety of equipment involved, along with other factors combine to make the inspection of almost any roadway project a challenge.

Whether the project involved is a highway, a long bridge, an airport runway, or rail construction, there is a horizontal construction factor to consider. The project is long and linear and there are certain challenges inherent in such a project which are not present to the same extent in vertical construction. On a project of any size, there must be multiple inspectors present at all times. If more than one activity is ongoing at any one time, odds are that no two are going on close enough to one another for one inspector to inspect both. Even if tasks are proximate, there are many activities, such as paving and pile driving, which demand both certification in the discipline and the undivided attention of one or more inspectors. Also, the contractor could have multiple crews, separated by great distances, performing one activity or item of work Commonly, activities are going on miles apart at the same time. Proper inspection under these conditions requires coordination and manpower.

The relative size, mobility, and unique nature of the equipment required for highway construction present another concern when planning for the inspection of a highway project. Paving trains, scrapers, bulldozers, graders, rollers, milling machines and various large trucks can all be moving at the same time, many times within a few hundred feet of each other, all within a narrow right-of-way. Many of the activities performed by this equipment require an inspector. All of these machines demand respect from a safety perspective. The subject of accidents on roadway construction projects is a theme for another report. Suffice to say that to inspect any one highway construction item safely, an inspector must be aware of all activities in progress, as well as how the public is flowing through the work zone.

Maintenance of traffic (MOT) is probably the one concern which most clearly separates highway construction from most vertical construction and other types of horizontal construction. It is also one of the major challenges facing State Highway

---

[1] Current Position: Instructor, Department of Civil and Environmental Engineering, The Pennsylvania State University, University Park, PA 16802.

Former Position: Chief Area Construction Engineer, Florida Department of Transportation, Tallahassee, FL 32399.

Agencies (SHA) and large municipalities. Safety for inspectors and contractor personnel, as well as safety for motorists is always a concern As SHAs and other owners wrestle with this immense problem, inspectors are charged with day and night inspection of MOT devices.

Not only must the configuration of the devices meet the contract documents, Traffic Control Plan (TCP), but the inspector must first make sure that the TCP meets requirements set forth by the Manual of Uniform Traffic Control Devices (MUTCD). The MUTCD is a publication of the Federal Highway Administration (FHWA). Adherence to this publication is required by law in many states. Also a concern for the inspector is whether or not the devices themselves are approved products and whether or not the devices are made of approved materials. The challenge of keeping abreast of the MOT on a project is a function of the size of the project, the number of and length of areas where work is going on and the proximity to the traveling public at any one time. On some projects, MOT is a twenty-four-hour-a-day job for two or three people per shift.

IN-HOUSE INSPECTION

The traditional way for an SHA to inspect construction projects is with in-house inspection. That is, using Agency employees who perform this function on a full time basis. There are numerous advantages to this method. Foremost of these is uniformity. When Agency employees are doing the inspection, the interpretation of standards, specifications and procedures is more uniform from job to job and from one geographic area to another.

A perceived advantage to in-house inspection is higher quality. This can be a topic of heated debate. The Quality Assurance Reviews of the Florida Department of Transportation (FDOT) reveal a disproportionate number of noncompliances on Consultant-inspected projects when compared to Department-inspected projects. This issue is revisited and discussed in length in a later section of this report.

Of course, there are numerous disadvantages also. These manifest themselves as advantages to other systems and are all explored in greater depth in subsequent sections of this report. They include limited flexibility, limited resources, limited knowledge in areas not in the "mainstream" of construction practice, and a possibility of having workers idle in slow periods.

CONSULTANT INSPECTION

One of the most common complaints heard from contractors regarding Consultant inspection is that it "creates another layer of bureaucracy". This, they assert, delays decisions, thus delaying their prosecution of the work. Reducing delays by reducing bureaucracy was one of the main reasons that Florida chose to decentralize the Department of Transportation.

Consultant inspection is an option open to any owner. It simply involves retaining a Consulting firm to perform inspection duties for the owner on a project basis or a general basis. The Scope of Services for the work of the consultant is written to best fit the needs of the owner. Duties may include the ability to make decisions of any type deemed necessary by the owner. The level of discretion given the Consultant as to how much money may be encumbered without owner-approval should be written in the Scope of Services. The consultant's duties may include materials testing, Quality Control, Quality Assurance, or anything else the owner desires.

Initially the complaint of "just more bureaucracy" was a valid one. Consultant Construction Engineering and Inspection (CEI) firms seemed insecure in their position, and some in the Department were not comfortable in having consultants making binding decisions for the Department. Therefore, many times, field level decisions were passed on to Department employees instead of being made by the consultant As time has passed, consultants have become more experienced, more capable, and more confident.

The Department, too, has changed. In recent years, the unofficial motto has been "let them do what they are paid to do". No more will upper-level consultant engineers need to consult a mid-level FDOT official for every field-level decision. The Department still employs Project Managers on the projects, but this person provides administrative expertise more than field management. Instead of calling the District Construction Office with a problem and asking how to solve it, the consultant is now more likely to call with a recommendation or a report of action already taken. This maturation process continues.

Training the consultant to do things "the SHA way" is important. Virtually any training available to Agency personnel must be made available to Consultant personnel. As mentioned earlier, this helps in the important areas of quality and uniformity. Several inspection functions require qualification or certification on the part of the inspector. In these cases, it is important that any training being offered toward qualification be offered to Consultant personnel in like numbers to Agency personnel, from the beginning if possible.

There is a distinct staffing advantage inherent in the use of consultants for CEI inspection. Proper project management dictates that a project be inspected with as few inspectors as possible. Therefore it is incumbent upon the Project Manager to commence operations as "lean" as possible. This is true whether the project is to be inspected by consultant or in-house personnel. As the project progresses, it often becomes necessary to bring in additional inspectors. This is accomplished easily enough with a consultant. The consultant simply hires another inspector or transfers one to the project. Most of the time, if the project is being inspected by in-house personnel, there is no inspector to move onto this project because all the other projects were also staffed "lean", and there is no one to spare. The staffing flexibility that goes along with the use of consultants is a source of great comfort to the project manager.

An advantage to Consultant inspection which is growing in significance is the ability to bring expertise in a certain area to the inspection team. If an owner has a project which requires expertise foreign to the owner, the owner can write in the Scope of Services that only firms with this type expertise need apply. This is much easier and more efficient than placing an ad in a trade journal for someone with this expertise, hiring them, and then paying them for the rest of their career, only to use them occasionally in that area.

The sudden upsurge in consultant use in Florida was a result of several factors. Without question, one of these was the sudden and significant increase in the Work Program in the late 1980s and early 1990s. This increase was accomplished despite a short period in 1988-89 when the Department experienced a funding shortage at the end of the fiscal year which left more in-house inspectors than projects to inspect.

In FY89-90, the total work program was $500,400,000 statewide. By FY 91-92, it had grown to $1,025,000,000. Therein lies one of the biggest advantages of thc use of consultant inspectors. It is much more desirable for an SHA to hire consulting firms during a boom and not hire them during a bust than to hire more employees for a boom and have them idle during a bust. Certainly, the prospect of reducing the number of consultant contracts during a bust is preferable to either of the alternatives faced by the SHA which hired for the boom: 1) Layoffs, 2) Personnel sitting idle.

There were other compelling reasons for the decision to so greatly increase the utilization of consultants. The political ramifications of privatization can not be ignored as a factor. This goes hand in hand with the move toward no-growth government and government reduction.

QUALITY AND COST: A COMPARISON

For Florida Department of Transportation construction projects completed in FY 93/94, the Construction Engineering, and Inspection (CEI) costs for the Consultant-inspected projects averaged $671,000, while the Department-inspected projects averaged $143,000.[1] This is a consultant-to-in-house ratio of 4.22:1. Consultant-inspected projects tend to be larger. For jobs let and awarded in FY93/94, the average Consultant-inspected project was built for $3,116,289, while the average Department-inspected project cost $966,406[2]. This is a ratio of 3.22:1.

TYPE INSPECTION	CEI COSTS	PROJECT AMOUNTS
Consultant	$671,000	$3,116,289
In-house	$143,000	$ 966,406
Consultant: In-house Ratio	**4.22:1**	**3.22:1**

[1] "Construction Engineering and Construction Analysis", Florida Department of Transportation.

[2] "Report on Construction Contracts, FY 93/94", Florida Department of Transportation.

This would seem to indicate that the cost of CEI inspection, as a percentage, rises as project costs rise, and/or that consultant inspection is more expensive than in-house inspection. A Department study tracking these costs does not bear out the former. Actually, as one would expect, the CEI costs, as a percentage, decrease as the size of the project increases. For jobs completed between June 30, 1992 and June 30, 1994 (FYs 92/93 and 93/94), the CEI cost as a percentage of the final contract amount was 37% for projects of $50,000 and under. The figure was 13% for projects of $20,000,000 and over. There was a consistent and mostly steady decline in the CEI cost as a percentage of the final contract amount as the project size increased.

CEI COSTS AS A PERCENTAGE OF FINAL CONTRACT AMOUNT

CATEGORY	FY92/93 & FY93/94 PERCENT	FY87/88& FY88/89 PERCENT
$50,000 and under	37%	31%
$50.000 - $249,999	21%	20%
$250,000 - $499,999	27%	16%
$500,000 - $2,499,999	19%	11%
$2,500,000 - $4,999,999	17%	11%
$5,000,000 - $9,999,999	18%	10%
$10,000,000 - 14,999,999	19%	11%
$15,000,000 - 19,999,999	12%	8%
$20,000,000 and over	13%	8%
Totals For All Categories	**17%**	**10%**

The obvious conclusion is that consultant CEI inspection is more expensive than in-house CEI inspection. This is borne out by the fact that this same study revealed that the average CEI cost as a percentage of total contract cost was 17%. A similar study done for projects completed between June 30, 1987 and June 30, 1989 showed that the average CEI cost as a percentage of the of total contract cost was 10%. This is significant due to the fact that the use of consultants as CEI inspectors rose so considerably during the interim between the two similar studies[1].

In the Quality Assurance Review performed in 1994 by the Office of Construction, there were 213 project reviews. Of these, 105 were inspected by consultants (49%). Projects inspected in-house numbered 108 (51%). There were 197 instances of non-compliance found. Non-compliance is defined as a process not in compliance with pre-determined criteria. Of these 197 instances, 125 were found on Consultant-inspected projects (63%). Seventy-two of these non-compliances were found on Department-inspected projects (37%).[2]

---

[1] The Florida Department of Transportation has no record of CEI cost as a percentage of contract cost for Consultant jobs only or for In-house jobs only.

[2] "Construction Quality Assurance Review Reports", Florida Department of Transportation.

Why do 49% of the projects experience 63% of the non-compliances? There are several factors which may explain this discrepancy. As mentioned above, the Consultant-inspected projects tend to be larger. Almost certainly this is a factor, since bigger projects usually have more activities and items where errors can be made. Another factor is the fact that the Department has done its own inspection for seventy-five years and utilized consultants to any extent for less than ten years. There is simply a learning curve that hasn't yet been completed. It has also been asserted that officials in the field do not sufficiently monitor the work of consultants, spending most of their time overseeing their own subordinates' work.

INSPECTION UNDER DESIGN/BUILD

Design-Build is a common practice in vertical construction. However, it is a concept not universally embraced by the highway and bridge construction community. The Florida Department of Transportation has moved ever so cautiously toward this practice in recent years. The Department's foray into this sort of system of project inspection came in the form of a $50,000,000.00 program funded by the legislature specifically for such projects. On these projects, the prime contractor was the "bidder" because of the contractor's superior bonding capacity. The contractor submitted a package which included a design consultant and a CEI consultant. This program was administered over a two year period, and many things were learned.

Foremost and practical was the concept that this "Design-Build" was a practice more suited for some types of projects than others. The practice seems to be best suited for large bridge projects, buildings, and rail corridors. The chief benefit of Design-Build is time savings. During the program, a 36% time savings was realized on average for the eleven projects which the program encompassed. There seems to be no measurable difference in price, although time is a big factor in real total cost. One reason the prices were so close to those under the traditional system could be that the Department continued to dictate to the firms so much of their methods and procedure. As Design-Build becomes a standard practice, the cost may well decrease as old processes are replaced with newer ones.

Innovation was not the big benefit many had hoped for. With one huge exception, none of the projects saw any innovation worthy of mention. This could also be because the Department's attitude and management were so identical to how they handled things traditionally that the firms handled themselves traditionally also.

Recently, FDOT awarded the largest Design-Build project in state history. Because this project called for such different contracting methods than had been employed heretofore, it was/is a very high profile job and required much effort in meeting all the complex challenges it presented.

Projects to be designed and inspected by consultants had traditionally been advertised and awarded to a consulting firm in the design stage, designed, and then readvertised and awarded to a consulting firm for CEI. At each step, proposals were received, graded, consultants shortlisted, and one ultimately chosen for the project, often after a formal presentation. Never did the same company prevail for both functions. For this project, design engineers and construction engineers examined the proposals and graded each independently of the others, for his/her area of expertise. The contractor ultimately chosen was the one deemed to have the best combination of design plan, construction plan, schedule, and price. A formula was used which added to the objectivity of the selection process. Published as a part of the Scope of Work and Services for the contract, the formula was as follows:

$$\textbf{Adjusted Score} = \frac{\textbf{Price Proposal - (MCT - PCT) * UDC}}{\textbf{Technical Score}}$$

Where : MCT = Maximum Contract Time; PCT = Proposed Contract Time; UDC = User Delay Costs per day[1]. From the formula, one can see that this is not only a Design-Build job, but also an "A+B" job, combining Design-Build and Time-Plus-Money bidding. A+B bidding is also a new concept for FDOT, which traditionally awards contracts to the lowest responsible bidder based on price only. The scope of services for this contract specifically states that "the intent of this design/build contract is to combine all work and services required for the project into one contract and point of responsibility". The Department was thwarted in its efforts to have the inspection as part of the package by federal funding restrictions. A consultant was retained by the Department for inspection purposes.

The contract also calls for the awardee to provide all supervision, labor, equipment, tools, supplies, permanent materials, and temporary materials required to:

1) Provide required geotechnical services

2) Properly perform the engineering design and review

3) Develop required construction plans and design load test plans

4) Execute the required construction and demolition work as specified

5) Provide the required documentation to the Department at the conclusion of the project, i.e., as-built quantities and drawings, etc.

This would have required any number of contracts between various sections within the Department and various consultants over a period of years. The obvious advantages to this method are the one point of contact for the Department (which will provide a DOT-employed project manager as the contractor's point of contact) and

---

[1] "Scope of Work and Services", Replacement of Blackwater Bay Bridges.

the speed with which decisions can be made when plan error, changed site conditions, and other potential points of contention become reality.

So great was the effort required on the part of the bidders, that $50,000 was awarded by the Department to each bidder to help cover their costs of producing a proper bid package. It is estimated that, on average, this covered about half of the cost of the package to the bidders. An encouraging aspect of the experience so far is the fact that the response from industry to the system has been very positive.

This contract was a big step for the Department. As stated earlier, FDOT desires to let a contract which allows the prime contractor to retain the design consultant and the inspecting (CEI) consultant. Of course, this method has disadvantages also. The most obvious is that if the designer, inspector, and contractor are "one", there is great potential for corruption. This is more than offset by the consulting firms' desire to continue doing business with the Department.

While at this time, there are no plans for a Warranty or a Maintenance requirement to accompany the design-build-inspect contract, the consultant selection process is deemed to be sufficient deterrent to an inspection consultant "looking the other way" when the contractor makes a mistake in construction. Consulting firms are chosen in Florida by proposal review and presentation - not low bid. A consulting firm who damages the Department continually by wrongly benefitting its partners in a contract will damage its own reputation severely - possibly irreparably. They might find further work with the Department difficult to come by.

The Department has taken a bold step in integrating many innovative contracting practices into one project. The progress of this project will be monitored closely. Unfortunately, at this time, there is not enough data for the Department to further judge the merits of A/E Inspection and Contractor Inspection on Highway/Transportation Construction projects.

A/E FIRM INSPECTION

As noted earlier, the Florida Department of Transportation has never had the same firm that designed a project perform inspection on that project. The Department would very much like to do so, but has thus far been unable to convince federal officials that this is a prudent inspection method.

CONTRACTOR INSPECTION

Though thwarted thus far in its efforts to allow the contractor to fully inspect its own construction, the Department has long required contractors to generate their own Quality Control (QC) Plan for different items in the contract. For years, contractors doing concrete work have had to submit a QC Plan. This program requires the contractor to perform his own tests on the concrete, and to document results and any changes made to correct unsatisfactory results. The Department then performs periodic Quality Assurance Reviews (QAR), to determine whether the process is satisfactory.

Therein lies the difference between QC and QA. Quality Control examines a procedure as it relates to a specific test or project. Quality Assurance examines a procedure as it relates to the process. To become a QA problem, a discrepancy must be systematic (found repeatedly). A single discrepancy is a QC problem, requiring immediate attention.

The Department is well along toward expanding this requirement of a QC Plan from the contractor to other areas of construction, such as fill. These requirements are made at this time through the State Specification Book, Special Provisions and other contract documents. The short-term goal is to get all of these requirements into the State Specification Book, thus standardizing these areas as concrete is now standardized.

QUALITY ASSURANCE IN THE STATE HIGHWAY AGENCY

It is hard to imagine a State Highway Agency in the mid-1990's without an active Quality Assurance Program. In Florida, Quality Assurance is the law. A Quality Assurance Review of the Construction Section of the SHA is different than a Review of any other section within the SHA. There is a tangible product to inspect and review. This is helpful in some respects and challenging in others. Since the product is tangible, it is not sufficient to do a paper review. The product itself can be and must be tested and inspected. The testers and inspectors must then be reviewed. This review of the testers and inspectors is the Quality Assurance Review and it takes time, personnel, and money.

The first step to a good review is to devote sufficient time, solicit sufficient input, and put due emphasis into choosing the areas of review. To do a thorough review, an SHA must procure or develop individuals knowledgeable in the areas to be reviewed. It has proven advantageous in the decentralized Department of a large state to perform the review of each review item in one district within a short time frame, combine the reports of each reviewer into one large report, and present the report to that district before turning attention to the next district.

There are logistical challenges to following this procedure. The challenges are greater in a large state with several districts. However, in this way, all reviewers are focused on the district at hand and everything is fresh on everybody's mind at the time of the closeout meeting or exit interview. Establishing a schedule for the year, spacing the reviews sufficiently, and granting all parties sufficient notice of the schedule are the keys to making the process work. This method has also proven to have less of a disruptive influence on normal district activities.

As a minimum, the Quality Assurance Review Report should focus on the following:

Central Office Action Item(s) - A task that the reviewer believes that the Central Office should execute in order to assist the District in complying with existing requirements or to alter existing requirements, thus improving the constructed product.

Corrective Recommendation(s) to the District - A recommendation to the District from the reviewer as to how to remedy a Non-Compliance.

Critical Requirement(s) - A measurable activity in an SHA process that indicates whether or not the process is being properly carried out and if the quality of the product is being controlled.

Finding(s) - A discovery by the reviewer in the course of the review . The "Finding" may be positive, negative, or indifferent. A discovery shall not be labeled a "Finding" if the discovery indicates that a process is not in compliance with predetermined criteria unless the violation is isolated. An isolated violation of such predetermined criteria may be a "Finding", as may an activity noted by the reviewer which is deemed poor construction practice or unsafe, but is not in violation of such predetermined criteria.

Need(s) to Improve Process - A recommendation from the reviewer to the District as to how to improve upon a negative "Finding".

Noncompliance Area(s) - An area where critical requirements indicate a process is not in compliance with predetermined criteria.

Outstanding Quality Area(s) - An area of unique superior performance due to magnitude of effort or innovation in concept or application.

Once the Quality Assurance Review Report is written, and the closeout meeting has been accomplished with the district, it is recommended that the report be entered into a "Quality Assurance Database" of some kind. This is a mainframe system for the official files of Quality Assurance information. The database should contain schedules, report summaries, and follow-up information for each Quality Assurance Review. This system is mainly for the use of the highest levels of management. However, the system is also of great use during follow-up reviews.

It should be noted that the review is on the construction process - not simply on the Construction Section - of the SHA. Therefore, it is essential that officials who make decisions which affect areas that impact the construction process not only get copies of the report, but also are present at closeout meetings. If this does not happen, the effectiveness of the program will surely suffer.

SUMMARY

There is no "best" way to inspect a transportation construction project. Each organization (owner) must make the decision as to what is best for them at any one time. What is best for an owner now may not be what is best for that owner two years hence. The factors that weigh in on this decision are many and diverse.

Political considerations play a huge role, as do changes in the size of work programs, consistency in the size of work programs, frequency of the need for the inspection of "specialty work", simple economics, and a host of others, large and small.

There is no doubt that the use of consultant CEI firms is more expensive in the short term, and letting the contractor inspect its own work or hire its own inspection may be somewhat risky. It is up to the management of each individual owner to decide if this short term expense or risk is justified by long-term benefits.

CONSTRUCTION INSPECTION IN THE WATER AND WASTEWATER INDUSTRY

W. Laird Ellis, Jr.,[1] Member, ASCE

INTRODUCTION

In an atmosphere of increasing concern over construction liability, architect-engineers and special consultants no longer "inspect" construction. They provide general administration of the construction contract on the owner's behalf and observation of construction to provide the owner a greater degree of confidence that the contractor's completed work will conform generally to the construction contract documents. These construction services include full or part-time observation of construction activities to determine if the work is generally proceeding in accordance with the contract documents, to identify unsatisfactory, faulty, or defective work, and to verify that tests, equipment and systems startups, and operation and maintenance training are conducted in accordance with the contract documents. Since the contractor is the only person who controls the means, methods, and materials to complete the construction project, he bears the authority and the responsibility for supervising construction operations and for safety on the construction site. For the purposes of this paper, architect-engineer means the designer of record, special consultant means a third-party firm independent of the architect-engineer and the owner, and inspection means administration of the construction contract and observation of construction activities.

Construction services, including observation of construction activities, identification of unsatisfactory work, and verification of compliance with construction contract documents, are needed on all construction projects to achieve a quality product. These services are typically provided by either the architect-engineer, the owner, or a special consultant. Some owners are beginning to require contractors to develop and maintain QA/QC programs, but contractors typically do not provide construction services. Water and wastewater construction can be divided into two areas of specialization, treatment plant construction and pipeline system construction for water distribution and wastewater collection. The following discussion of the three methods for providing construction inspection is presented in terms of plant construction and pipeline system construction.

WATER AND WASTEWATER TREATMENT PLANT CONSTRUCTION

Water and wastewater treatment plants are typically designed by architect-engineers engaged by the municipality or regional agency which will operate the plant upon completion of construction. As with other fields of construction, treatment plant construction involves specialized means, methods, and materials that have been

[1] Senior Engineer, Turner Collie & Braden Inc., Dallas, Texas.

developed to provide durable, stable structures and reliable, efficient operations. While many general construction techniques are applicable to plant construction, specialized techniques are used for installation of equipment, construction of hydraulic structures, and other treatment plant construction activities. Since the purposes of observing construction include determining compliance with contract documents and identifying defective work, the construction inspector must be able to recognize work that has been performed correctly. This ability requires knowledge and experience that can only be gained through participation in numerous treatment plant construction projects. Increasingly complex treatment processes and equipment are needed to meet rapidly changing performance requirements for both water and wastewater treatment plants. Because of the high capital costs and the technical complexities associated with treatment plants, most owners prefer to construct new facilities only when forced by increased population or regulatory edict. This tends to make treatment plant construction an infrequent event for most owners. Architect-engineers participate in numerous treatment plant construction projects by designing plants for numerous owners.

Architect-Engineer

Architect-engineer inspection for water and wastewater treatment plant construction has historically been the method of choice for observing construction activities. Because the architect-engineer developed the treatment process and prepared the design, the designer's professional staff is most familiar with the construction documents and the intent of the design. The designer has incorporated lessons learned through numerous projects into the contract documents. Through repeated projects, the designer's staff has developed considerable expertise with specialized treatment plant construction. Design consultants are able to maintain a staff of plant construction experts in part because their engineering practice requires the ready availability of people knowledgeable in plant construction. Under one method for reducing the cost of inspection, the designer can provide a highly experienced representative to direct the observation of construction using the owner's in-house staff. The designer's representative provides the specialized construction expertise while the owner's existing staff supplies the manpower. This solution can reduce the owner's cost of construction phase services while still providing the necessary expertise to ensure a quality project. When the designer's representative supervises the owner's staff, the contract between the owner and the designer must be written so that it spells out the professional's responsibility and the relationship with the owner's staff.

Owner

Owner inspection of water and wastewater treatment plant construction refers to situations where the owner is not the architect-engineer. Because typical owners operate only one water treatment plant and one wastewater treatment plant, their staff can only experience treatment plant construction when those plants are modified or expanded. Even in larger operations where additional treatment plants are needed, the number of plants is kept to a minimum. When treatment plant construction occurs infrequently, owner personnel have little opportunity to become familiar with the specialized treatment plant construction practices and to gain adequate experience to inspect treatment plant construction. Without involving the architect-engineer, the owner may not have the familiarity with and understanding of the project design needed to interpret the construction documents. Access to the knowledge of how layouts were made or why materials were chosen or how field changes will affect the treatment process is essential to the construction of an efficiently operating treatment plant. Without intimate involvement in the design process, the owner cannot provide this knowledge as effectively as the architect-engineer.

Special Consultant

Situations occur which require the use of special consultant construction inspection. Special consultants bring their knowledge of plant design and construction practices to the project. This can include knowledge of specialized or innovative techniques not available from either the designer's or owner's staffs. A special consultant is typically brought into the project after design has been completed in order to observe the construction process. Because the special consultant did not prepare the construction contract documents, requests for clarifications or changes to the construction contract documents can be referred to the designer for evaluation or to the owner for approval. The special consultant observes construction activities to determine if the work is in accordance with the contract documents.

Summary

The most important aspect of construction inspection is to bring the greatest knowledge and experience to the project. This helps assure the owner of a quality project. The architect-engineer should administer the construction contract and visit the site on a regular basis. The continued involvement of the team that prepared the design helps minimize field changes that could negatively affect the completed project and helps develop the appropriate response to changed conditions in the field. Observation of the construction process should be performed by individuals knowledgeable in the type of construction being performed. These "inspectors" can be provided by the architect-engineer, the owner, or a special consultant. With the architect-engineer administering and interpreting the construction contract and knowledgeable inspectors observing the work, the owner can be more confident the completed work conforms to the construction documents.

WATER DISTRIBUTION AND WASTEWATER COLLECTION SYSTEM CONSTRUCTION

Water distribution and wastewater collection systems spread throughout the owner's service territories. Water distribution system facilities include pumping stations, piping, valves, fire hydrants, and ground and elevated storage tanks. Wastewater collection systems consist of sewers, manholes, lift stations, force mains, and other facilities. Because of frequent minor modifications and additions to their systems, many owners maintain a professional staff to prepare construction contract documents for minor system expansions and modifications. Outside consultants are typically engaged for projects too large or too specialized for the owner's staff.

Architect-Engineer

When collection and distribution facilities are designed, architect-engineer can refer to two very different organizations. The first is the independent design consultant whose professional engineering practice specializes in water and wastewater facilities of all types. The second organization consists of the in-house professional engineering designers employed by the owner. This owner is typically a large municipality or regional agency with a need for design services on a routine basis. When the construction project involves an in-house design, owner inspection of the construction project includes all of the advantages of architect-engineer inspection. At the end of the design of a water distribution or wastewater collection system project, no one will be more familiar with the project and its associated construction contract documents than the designer. The designer knows what facilities will meet the owner's needs, how the design was developed, and why the specific layout or route was chosen. This familiarity with the project and with the design, coupled with an intimate knowledge of the contract documents, is the strongest recommendation for construction observation by the designer.

Owner

Owner inspection of water distribution or wastewater collection system construction projects refers to projects where the owner's staff did not prepare the design. There are several reasons why owner inspection is an acceptable option for system construction. Since the collection and distribution systems comprise the majority of both the water and wastewater system facilities, a typical owner's staff deals with system construction on a regular basis. There are many more opportunities for staff members to develop the in-depth knowledge needed to inspect construction. This experience means that contractors will employ construction means, methods, and materials that are familiar to owner personnel. In addition, the owner's staff will possess the greatest familiarity with the existing distribution and collection systems. When construction decisions need to be made that affect the operation of the distribution or collection system, the owner's staff will be best prepared to address the issues. For these reasons, owner inspection of water distribution and wastewater collection system construction can be an acceptable alternative. When the architect-engineer does not provide continuous construction observation, the designer can still administer the construction contract and make periodic site visits in support of the owner's inspectors.

Special Consultant

For the same reasons that a special consultant can contribute to the inspection of treatment plant construction, special consultant inspection can contribute to distribution and collection system construction. A special consultant comes to the project with extensive experience in the types of construction utilized in the work. Since the special consultant possesses less familiarity with the construction contract documents and the design intent than the architect-engineer, the designer can be retained to help interpret the construction contract documents during construction. The special consultant will observe construction activities to determine if the work is in accordance with the contract documents.

Summary

The most important aspect of inspection is to bring the greatest knowledge and experience to the construction project. The architect-engineer, whether a consultant or the owner's employee, should administer the construction contract and visit the site on a regular basis. Observation of the construction process must be performed by individuals knowledgeable in the types of construction being performed. These inspectors can be provided by the designer, the owner, or a third party. The owner can be more confident the completed work conforms to the construction documents with effective administration and interpretation of the construction contract and knowledgeable observation of the work.

CONCLUSIONS

Historically, inspection of water and wastewater construction has been performed by the architect-engineer. In recent years, many owners have elected to use their own personnel to inspect construction activities. Third party inspection by independent consultants is also on the rise.

Construction inspection by the architect-engineer, whether an independent consultant or the owner's in-house staff, brings the advantages of familiarity with the design, intimate knowledge of the project, and construction experience garnered on numerous projects. A concern expressed by some owners for designer inspection is that the designer has an interest in minimizing the identification of any errors or omissions that might be present in the design. This could lead to unethical behavior in an attempt to cover up the problem.

The owner's staff brings an intimate familiarity with the existing facilities to the construction project. When the owner has not prepared the construction contract documents, the decision to use owner inspection should be based on an evaluation of the project and the construction experience of the owner's staff. Several disadvantages relate to owner inspection of water and wastewater construction. When the owner does not prepare the construction contract documents, administering the contract can be difficult. The owner's staff will not have the designer's familiarity with the contract documents or the reasons for performing the work as specified or shown on the drawings. By allowing field changes without consulting the designer, the inspector can conceivably alter the project so that it will not perform as required. The special consultant brings an extensive knowledge of design practices and construction techniques to the construction project. While less familiar with the design, the construction contract documents, and the existing facilities and systems, the special consultant can supply knowledge of specialized or innovative construction techniques that may not be available from either the designer's or owner's staffs. The special consultant provides construction inspection experience when, for whatever reason, the designer and the owner cannot. With contract document interpretation assistance from the designer, a special consultant can help assure the owner of a completed project that conforms to the construction documents.

CONSTRUCTION INSPECTION FOR COMMERCIAL BUILDINGS

James H. Baldridge, P.E.,[1] Member, ASCE and Dan K. Williams, P.E.[2]

INTRODUCTION

The importance of construction inspection is undeniable. One need not look far to find examples of where a commercial building project's inspection program failed and a catastrophic loss resulted. These examples run the gamut from the large, nationally publicized losses, to the neighborhood newspaper's second page variety. What is common to each is that there was probably an inspection program that could have and should have prevented the loss. This paper will introduce and discuss some of the issues related to "Who" provides commercial building construction inspection. The foremost intent of this paper is to improve the quality of the constructed project.

Quality assurance and quality control (QA/QC) programs are necessary and required to help assure that a commercial building construction project provides the owner with a completed building of good quality, and protects the public health, safety and welfare during the buildings use and occupancy after completion.

Four (4) methods of construction inspection for commercial buildings will be presented and discussed. A definition of "commercial building" will be presented and described in terms of the types and sizes of facilities that fall within the context of commercial buildings. This will be followed with a presentation and description of the types of inspections, e.g., building, mechanical, electrical, etc.. Finally, the key participants of the building process will be identified and discussed as they contribute to and/or relate to the methods of inspection of a commercial building project. Each of the four (4) methods of inspection will be discussed in the context of its advantages and disadvantages.

At the outset it must be understood that there is generally no individual method, or technique, that is clearly better than another. On the other hand, for a particular set of circumstances, there will probably be a reasonable number of alternatives that could be considered for selection of the "best" alternative for a commercial building project. When there are alternatives, the determination of "best" approach, as is true with any other determination or choice on a construction project, is left to the owner with advice from the project team. It is hoped that the following discussion will be helpful in making decisions about inspection services.

[1] Principal Engineer, Engineering Consulting Services, LTD., Chantilly, VA 22021.

[2] Supervisor of Critical Structures Section, Department of Environmental Management, County of Fairfax, VA 22035.

DEFINITION OF COMMERCIAL BUILDING

There may be as many definitions of the term "commercial building" as there are people to define it. Building codes often differentiate between "residential" buildings and a variety of other building types (factory, assembly, institutional, storage, etc.), all of which have particular code requirements. Usually, building types are defined in terms of "Use Groups" for the purpose of articulating fire protection and structural requirements. This paper will focus on commercial buildings only and will discuss commercial building inspection programs.

More and more, as updates to building codes occur, code language will associate commercial building inspections with the term "Special Inspections". This paper will use this terminology as well. Generally, inspection programs for commercial buildings are the focus of and within the jurisdictional province of federal, state or local regulatory agencies; therefore, the following discussion will have some nuance of code language.

Finally, there will be no distinction between inspection of new construction and that which is provided for renovation or addition to existing facilities.

TYPES OF COMMERCIAL BUILDING INSPECTIONS

A thorough discussion of "who should provide commercial building inspections" requires us to introduce the topic of commercial building "systems". Commercial buildings are assembled from systems, i.e., foundation, structural frame, mechanical, electrical, plumbing, EIFS, fire protection, etc. A commercial building project that requires construction inspection may simply be the upgrading or replacement of a mechanical system.

For example, a presently on-going commercial project is located at the Pentagon. The heating plant for the world's largest office building involves the replacement and upgrading of the building's heating system. The inspection objectives are to facilitate the highest quality work, consistent with the project's approved construction documents and the A/E's design intent. These objectives are consistent with those for a new high-rise commercial office building. Only the scope of the program and the expertise of the inspectors differ.

Commercial building inspection is a wide and varied field of knowledge on which volumes have been written (currently the US Army Corps of Engineers has a four volume set of handbooks that it uses for construction inspection), but the intent and objectives of the activity are very narrowly focused on providing quality. The team that facilitates this activity is a function of the activity, and it is the type of activity that will help define "who" provides it.

The following types of commercial building systems will generally produce construction activity that will require inspection:

1. Building: There are a number of components of this system that constitute the total construction activity. Each has a unique field of knowledge. These components are often specified for "independent" or "third party" inspection and frequently require specific "certification" of the inspectors, i.e., ACI Inspector, Level I certification, AWS Certified Welding Inspector, etc.

The components of the Building system are any combination of the following: Steel, Concrete, Masonry, Wood, and Soils.

2. Mechanical, Electrical, Plumbing (MEP): The components associated with this system are highly specialized and have unique fields of knowledge. But, unlike the Building system components listed above, the inspection activity associated with this system is usually retained in and reserved by the building code official for his own staff.

3. Fire Protection: Again, like MEP systems, fire protection systems are specialized and very technical. Generally the associated inspection activity is retained by the code official for highly trained individuals on his staff.

The above discussion notwithstanding, any one of a commercial building's systems that has inspection activity required by the construction documents or the applicable building code can effectively be inspected by any one (or combination of more than one) of the inspection methods that follow.

Ultimately, the successful method chosen by the owner to implement the discretionary components of a commercial building's inspection program is the one that will facilitate the highest quality of construction and be consistent with the approved construction documents and the design intent.

(Note: Testing laboratories are not discussed in this paper. They are unique organizational entities that augment the inspection activity regardless of method chosen. The laboratories that support commercial inspection programs are usually accredited by nationally recognized agencies and are carefully regulated. It is left to the discretion of the owner to select one that has the proper accreditation for the inspection activity being considered.)

METHODS OF CONDUCTING INSPECTIONS

Before discussing the various methods for performing commercial building inspection, it is important to note that with the exception of Federal Government projects, the "authority" for private sector compliance inspection on commercial buildings almost always resides with the State.

As a general rule, State Law assigns the Code Official the responsibility for compliance inspection. The genesis of this authority probably comes from the need to have a single source of responsibility for public safety for the constructed project. Usually, and in turn, the Code Official may delegate this responsibility for compliance inspection to approved individuals and agencies. The following paragraphs will describe the methods, or categories of individuals and agencies, that can provide compliance inspection services. As each method is discussed, its advantages and disadvantages will be presented.

Note: The arguments surrounding the issue of full time versus part time inspection strategies are left to the owner for benefit-cost analysis and other forms of resolution. It should be noted that this issue is often not discretionary for inspection activities related to some components of Building systems (i.e., structural concrete placements, controlled engineered fills, etc.), since applicable building codes may dictate the strategy. For other systems and components, however, the analysis must often address and rationalize the risk of not achieving full compliance with construction documents and design intent.

Owner Inspection Team

This method of inspection is usually implemented by large corporate developers, owners and government agencies that retain full-time staff for the purpose of providing inspection for not only new construction projects but also for the day to day "facilities" type of inspections required to operate and maintain property portfolios.

This is arguably the most effective method in terms of insuring the quality of the project and producing compliance with the approved construction documents. The team would have the greatest familiarity with the owner's intent and would be the most diligent in seeking compliance so as to mitigate potential future liability.

This method may, however, be the most costly. In many, if not most instances, the frequency of commercial construction projects is not great enough to justify keeping a full time inspection staff on the payroll. Therefore, some of the benefits that accrue to an owner who keeps a full time inspection staff are lost when he hires in-house staff for transient needs. In this situation other methods of commercial inspection may be more desirable.

The A/E Inspection Team

This method of inspection is intuitively the one that will provide the greatest likelihood of compliance with the approved construction documents. It is also the one that will best facilitate coordination between the construction contractor and the design professional.

A disadvantage associated with an A/E inspection team is the reluctance of the A/E to diligently pursue compliance with the construction documents when there is a design error that was not corrected before the project entered construction. The A/E may deal with such an error unfairly and provoke construction claims and/or litigation by the building contractor. Cost is another disadvantage to this method, in that the designer frequently does not have inspection staff and has to hire thus losing some of the intuitiveness benefit of the argument stated above.

Contractor Inspection Team

For many years, this method has been widely used in Federal government contracting, often being referred to as Contractor Quality Control (CQC). In much of the private sector, this method of building inspection is prohibited because of building code requirements. It appears that the former approach is waning, and the latter is on the rise.

The arguments supporting this method are exclusively cost driven with the perception being that the contractor is most familiar with the means and methods of construction and is entirely in control of the construction schedule. Given those attributes, there is little doubt that the most cost efficient inspection team would be controlled by the construction contractor.

On the other hand, this same construction contractor usually works in a market where the lowest bidder gets the job. Qualification based selection has yet to penetrate the commercial building sector of the construction industry. This inspection team would be under pressure to perform inspections so as to produce as little expense as possible, and it is generally not in the best interest of the owner or the public to have this constraint govern commercial building inspection activity.

Finally, and again relying on intuition, there is little justification for a compliance inspection effort being managed by the constructor.

Independent Consultant Team

A third-party independent consultant offers an inspection approach that not only provides the opportunity to involve a design professional in the commercial building inspection process, but offers a highly reliable technique to facilitate compliance with the approved construction documents.

Many building code officials are now requiring owners to hire independent third party inspectors to perform compliance inspections for building systems. This option has been in the building codes for several years, but now as government budgets are being pressured to shrink, third party inspection is being viewed as an effective way to cut expenses without compromising public safety. State and local jurisdictions are not only placing more reliance on third party inspection services offered by registered professionals for Building systems but are also considering this technique for the Fire Protection and MEP systems inspections, as well.

The "Intimate knowledge" and timely coordination disadvantages are similar to other methods that do not have the designer performing the inspections, though costs are most likely less.

CLOSING

A number of code officials are now requiring the owner to prepare a "schedule of special inspections" and submit it as part of the permit application. This schedule is offered by BOCA upon request and provides a worksheet that assists the owner in thinking through and developing a commercial building inspection program that includes the method(s) discussed above.

There are many efficient and effective combinations of the above four (4) methods that will produce a quality project that is consistent with the approved construction documents and design intent. The decision on which method to select for a given project is the owner's. It is only recommended that the selection be made after listening to the concerns of the project's design professionals, coordination with the appropriate code officials, and interviewing recommended and qualified inspection personnel when no in-house capability exists.

> Quality is never an accident. It is always the result of high intention, sincere effort, intelligent direction, and skillful execution. It represents the wise choice of many alternatives.
>
> Anonymous

CONSTRUCTION INSPECTION IN THE RESIDENTIAL BUILDING INDUSTRY

John M. Hunter, P.E.,[1] Member, ASCE

INTRODUCTION

Construction can be classified as residential, building, engineering, or industrial. Residential construction comprises about 30 to 35 percent of new construction during a typical year (Clough, R.H., 1986, *Construction Contracting,* Fifth Edition, John Wiley and Sons). Residential construction includes single family homes, duplexes, townhouses, condominiums, apartment buildings and public housing projects.

For the purpose of this paper, discussion is limited to inspection of single family homes, townhouses, and duplexes. Multifamily homes (e.g., apartment buildings, public housing projects) can be classified as commercial building construction.

Residential construction is unlike the other categories. In the building, engineering and industrial construction sectors of the construction industry; an owner normally retains a design professional to design the project and a contractor to build the project. This is the "traditional" or design-bid-build approach that is used extensively. Other project delivery system, including construction management-for-profit and design-build, are also used. Inspection is normally performed by the Owner, the design professional, construction manager (construction management-for-fee or agency construction management), or an inspection consultant.

In residential construction, there is usually no "design professional" per se, except for luxury type homes. Design is accomplished in different ways. Standard plans, or modifications thereof to suit the tastes and needs of a family, are often used for single family dwellings. The standard designs may be made available by builders or developers. Custom designs may be developed by the owner, possibly in conjunction with an architect or the builder. In the case of speculative construction, a developer develops a tract of land and builds houses of different designs for sale, or sells lots for development by builders or owners. The owner in this case is not a part of the "design phase", but buys a house that has already been constructed.

Single family homes are normally constructed by developers and builders, although some home owners may build all or part of the homes.

Inspection of new residential construction is performed primarily by public agencies. It is believed that inspection of residential construction started as a result of failures, problems with quality, and the so-called "human need to regulate". Private inspection

[1] Director, Construction Services Division, Herbert, Rowland & Grubic, Inc., Harrisburg, PA 17111.

agencies or the design professional, if one exists, may be used for inspection of luxury home construction. Lending institutions who provide money for residential construction mortgages inspect the work during various phases of construction to verify completion for partial payments.

Owning a home is the largest single investment that most families make. From the perspective of a family, quality and safety should be of considerable importance. Therefore, a degree of inspection is necessary to provide a reasonable degree of confidence that quality and safety standards are met. Inspection by various entities is discussed below.

Inspection by Government Agency

Construction of new residential construction may be regulated by laws or local ordinances, usually in the form of building codes. Inspections are conducted by government agency personnel, such as code enforcement inspectors. Prior to construction, the owner or builder may be required to submit plans to the responsible government agency (e.g., code enforcement office) for review. If the plans are acceptable, a building permit is issued. Specific phases of construction are inspected by agency inspectors. Typical elements for inspection are the foundation, structural framing, plumbing and electrical systems.

The plans and construction must conform to code requirements. An acceptable code for residential housing construction is the *One and Two Family Residential Code* published by the Council of American Building Officials (CABO). Some building officials use the *Basic/National Building Code* published by the Building Officials and Code Administrators International (BOCA) for residential construction.

If codes other than CABO code are used, they should be reviewed and excessive or unnecessary requirements should be eliminated. Codes must have flexibility so that changes which do not sacrifice safety or quality can be made easily.

For the specific case of single-family, townhouses, and duplex residential housing construction, it is believed that codes enforcement inspection of critical elements of a house is adequate for health and safety purposes. Inspections beyond those required by building codes and local ordinances will add to the cost of a new home and may not significantly improve quality or make the home any safer. Any inspections above and beyond those required by building codes should be voluntary and at the homeowner's discretion.

Inspection by Owner

Technical inspections should be conducted by code officials or qualified individuals. For those home owners who have a contractual agreement with a builder or developer for construction of a new home, the homeowner's "inspections" should be limited to the aesthetic and functional characteristics of the dwelling as they pertain to the contractual agreement between the parties. The typical homeowner does not have the expertise to inspect the technical aspects, for example structural framing, of a structure. If the owner wants inspection beyond code inspections, then the owner can contract with a qualified home inspector or consultant.

Inspection by Consultant

A homeowner may retain a consultant to inspect the construction. This inspection would be in addition to the inspections performed by governmental agencies, and would add to the cost of the home.

Inspection by Design Professional

An owner who retained an architect to design a luxury home may retain the architect to inspect the work to verify compliance with plans and specifications requirements.

RECOMMENDATIONS

The following recommendations are offered for inspection of single-family homes, townhouses, and duplexes:

- inspections should be limited to those performed by local government agencies, such as code enforcement offices,
- inspections in excess of those performed by code enforcement inspectors should be voluntary and at the owner's option,
- an appropriate code, such as the CABO *One and Two Family Residential Code,* should be used,
- codes should not have excessive or unnecessary requirements,
- codes should be flexible so that changes which do not affect safety or quality can easily be made; and
- specific phases of the work that require inspection before proceeding to the next phase should be established.

CONSTRUCTION INSPECTION IN THE WASTE MANAGEMENT INDUSTRY: MUNICIPAL SOLID WASTE FACILITIES

John M. Hunter, P.E.,[1] Member, ASCE

INTRODUCTION

"Who provides construction inspection" in the waste management industry is largely determined by federal and state laws and regulations. Waste management facilities are unlike other types of construction. Waste management facilities are heavily regulated by the federal and state governments in order to protect human health and the environment and to address the concerns and perceptions of the public. The intent is to give the public confidence in the safety and performance of waste facilities. Failures of waste management facilities could result in contamination of the environment in general and ground water in particular. Contamination could render the ground water useless or may involve costly cleanup.

The term "construction inspection" is not generally used in waste management facility construction. Rather, the term "construction quality assurance (CQA)" has come in to use. CQA includes construction observation or monitoring as well as testing and documentation. Some CQA organizations use the term "construction monitor" in lieu of construction inspector.

This paper addresses municipal solid waste facilities only. Those facilities include landfills, surface impoundments and waste piles. Landfills are the most common of the types of facilities constructed. The focus of this paper is on municipal solid waste landfills. Hazardous waste storage and disposal facilities and remediation projects are beyond the scope of this paper.

DEFINITIONS

The control and assurance of quality materials and workmanship involves construction and manufacturing quality control and construction and manufacturing quality assurance. Brief definitions are:

> *Manufacturing Quality Control (MQC):* A planned system of inspection that is used directly to monitor and control the manufacture of a material which is factory originated.

---

[1] Director, Construction Services Division, Herbert, Rowland & Grubic, Inc., Harrisburg, PA 17111.

Manufacturing Quality Assurance (MQA): A planned system of activities that provide assurance that the materials were manufactured as specified.

Construction Quality Control (CQC): A planned system of inspections that are used to directly monitor and control the quality of a construction project.

Construction Quality Assurance (CQA): A planned system of activities that provide assurance that the facility was constructed as specified in the design.

These definitions apply to both synthetic and natural materials and related workmanship.

BACKGROUND

Subtitle D of the Federal Resource Conservation and Recovery Act (RCRA) establishes minimum requirements for the waste management industry. States are required to submit plans to the U.S. Environmental Protection Agency (EPA) for approval. The states must show how they are going to develop and implement permit programs for waste management facilities. Federal regulations include minimum design criteria which the states must meet or exceed. Waste management facilities are permitted by state governments. A permit establishes requirements for design, construction, operation, and closure of the facility. The landfill owner/operator is the permittee and is responsible for compliance.

State regulations may include design requirements, requirements for CQA plans, and certification of completed construction facilities. CQA plans include procedures for observing sampling, measuring, testing and documenting facility construction activities; and cover personnel qualifications. Although the owner is responsible for setting up the CQA program and retaining the services of a CQA firm, the CQA firm must remain independent of the owner. This is necessary to avoid conflicts of interest and to remove perceptions of impropriety.

There are a number of organizations involved in a waste management project. Included are the owner/operator (permittee), design professional, construction/manufacturing quality assurance organizations, surveyor, testing laboratory, manufacturers, suppliers, installers, contractors, host municipality, and regulators.

CQA of waste management facilities requires specialized expertise. Personnel qualifications have been recommended. For example, the EPA Technical Guidance Document titled "Quality Assurance and Quality Control" (EPA/600/R-93/182) contains recommended minimum qualifications for CQA, CQC, MQA and MQC personnel.

The National Institute for the Certification of Engineering Technicians (NICET) has prepared an inspection certification program for geosynthetic materials installers. The program includes nine subfields. The EPA Quality Assurance and Quality Control technical guidance document makes reference to the certification program. Construction monitors on waste management facilities may hold NICET soils certifications and be certified nuclear density gage operators.

Owner

The owner is responsible for establishing a CQA program and is ultimately responsible for compliance with the CQA program approved under the permit. The owner may design the project and prepare a CQA plan. Although the Owner sets up the CQA program and retains a CQA firm to implement the plan, the CQA organization must remain independent. The CQA organization must be unobstructed in the performance of its duties and must not be influenced by the owner.

Design Professional

CQA may be performed by the design professional retained by the owner to design the project and prepare the permit application. Some specialist design professionals not only have construction observation personnel, but can also perform materials testing. Other design professionals only perform construction observation and subcontract materials testing or coordinate the work of laboratories under direct contact to the Owner for third-party quality assurance testing.

There are advantages to the design professional performing CQA/MQA. The design professional is a specialist in waste management engineering and as the designer of the project has an intimate understanding of the assumptions used in developing the design as well as an understanding of the site conditions. Field problems or questions that arise can be handled quickly. The design professional's personnel would be on the site communicating directly with personnel at the home office.

One of the disadvantages cited to this approach is that the design professional may want to cover design errors and omissions during construction. The design professional has an obligation to the client to provide services based on a relationship of trust and confidence. The best interests of the design professional's client include advising the client of problems encountered and correcting those problems in a professional manner.

The design professional is normally the certifying engineer. Certification reports include, test results, record drawings, discussion on deviations and nonconformities, and other relevant information; and are submitted to the regulatory agency.

Contractor

Contractors, manufacturers and installers cannot provide CQA/MQA since they are the organization performing the work. CQA must be performed by an independent third party. However, contractors, manufacturers and installers do provide quality control. Manufacturing and construction quality control were defined previously. CQC/MQC requirements and procedures may be incorporated in the approved CQA plan. CQC/MQC is normally assigned to contractors, manufacturers and installers by contracts between the owner and contractors, manufacturers and installers.

Manufacturers, contractors and installers are required to provide test results and observations to the CQA organization and the information is incorporated in the project documentation and becomes part of the permanent record for the project. CQC/MQC is in the best interests of the contractors, manufacturers and installers because it requires them to be involved in the quality control process and to take proactive measures to assure materials and workmanship quality.

Independent Consultant

An owner may retain an independent professional or specialized consultant to provide CQA. This approach satisfies the concern of some that the design professional should not be inspecting its work. This approach provides for true third-party independence. However, the independent consultant, although probably a specialist in waste management construction, is not the designer and may not fully be aware of the assumptions that have been incorporated in the design as well as the detailed history of the design and problems encountered.

Problems or questions that arise on site are not as readily addressed and solved by the independent consultant who must go back to the designer for review and guidance. This could result in site visits which could lead to delays in the performance of the project. There may be temptation for the independent consultant to solve the problem or make changes without first consulting the design professional. This could lead to problems later and raises issues such as professional ethics and legal liability for changing the design of another.

The independent consultant may be the certifying engineer, fulfilling the certification services described under the "Design Professional Section".

Regulatory Agency

The work is subject to periodic site visits and inspections by regulatory agency personnel. Agency inspectors usually conduct an inspection after the agency has been notified by the permittee that a construction activity has been completed. Inspections can be unannounced. The inspectors check to verify that the facility is constructed in accordance with installation plans and design drawings. Reports are prepared and given to the permittee. During construction or at the end of construction, certification reports are submitted to the regulatory agency for review and approval. Some municipal waste landfill owners submit certification reports as layers are constructed to reduce the lag time in receiving approval from the regulatory agency. Waste cannot be placed in the facility until the agency has approved the facility.

SUMMARY

Waste management facilities are unlike other construction projects in that they are heavily regulated. The public is concerned about landfills and other waste management facilities. There are federal and state requirements for design, permitting and construction. The objective is to protect the environment and the public health and to instill a level of confidence in the public that the facilities are being properly designed, constructed and operated.

The “inspection” of waste management facilities’ construction involves an elaborate construction and manufacturing quality assurance and control program that involves a number of organizations. Unlike other types of construction, the Owner, who is responsible for providing a CQA program and who is ultimately responsible for compliance with permit requirements, cannot be directly involved in CQA. The CQA must be performed by an independent (third party) organization.

INDEX